the SCIENCE *library*

SPACE

the SCIENCE *library*

SPACE

John Farndon

Consultant: Sue Becklake

12·09

MC
PUBLISHERS

This 2009 edition published and distributed by:
Mason Crest Publishers Inc.
370 Reed Road, Broomall, Pennsylvania 19008
(866) MCP-BOOK (toll free)
www.masoncrest.com

Library of Congress Cataloging-in-Publication data is available

Space
ISBN 978-1-4222-1555-5

The Science Library - 10 Title Series
ISBN 978-1-4222-1546-3

Printed in the United States of America

First published in 2004 by Miles Kelly Publishing Ltd
Bardfield Centre Great Bardfield Essex CM7 4SL

Editorial Director Belinda Gallagher

Art Director Jo Brewer

Editor Jenni Rainford

Editorial Assistant Chloe Schroeter

Cover Design Simon Lee

Design Concept Debbie Meekcoms

Design Stonecastle Graphics

Consultant Sue Becklake

Indexer Hilary Bird

Reprographics Stephan Davis, Ian Paulyn

Production Manager Elizabeth Brunwin

www.factsforprojects.com

Contents

How to use this book

SPACE is packed with information, color photos, diagrams, illustrations and features to help you learn more about science. Do you know how far from the Sun Earth is or how many days it takes Pluto to orbit the Sun? Did you know that the biggest asteroid ever discovered measured 940 km across and that sunlight takes three minutes to reach Mercury. Enter the fascinating world of science and learn about why things happen, where things come from and how things work. Find out how to use this book and start your journey of scientific discovery.

Main image
Each topic is clearly illustrated. Some images are labeled, providing further information.

Main text
Each page begins with an introduction to the different subject areas.

The grid
The pages have a background grid. Pictures and captions sit on the grid and have unique co-ordinates. By using the grid references, you can move from page to page and find out more about related topics.

24

Glowing lights

LIKE THE Sun, stars are huge, fiery balls of incredibly hot gases. They shine because they are making energy. Deep inside every shining star, huge pressure squeezes atoms of hydrogen gas together to create nuclear reactions millions of times more powerful than a nuclear bomb. These reactions boost the star's core temperature, so much that the surface glows white-hot. A star continues to glow, sending out light, heat, radio waves and other radiation until all the hydrogen is used up.

1. A star is born when nuclear reactions begin

2. A star burns steadily

3. Dust swirling round a new star may form planets

4. A nebula formed from cloud and dust

▲ The four main stages of a star's formation.

● **Star quality**
Stars create their energy in the same way as nuclear bombs, but they rarely explode. Medium-sized stars burn steadily for billions of years because there is a balance – between the heat energy, which pushes gases outward as they expand, and gravity, which pulls them back in. Only when the star's nuclear fuel burns out, is this balance broken, and the star shrinks or, on some occasions, explodes.

◀ A medium-sized star.

● **A star's life**
Stars are starting up and dying do over the universe. They begin in gia of gas and dust, where material ga clumps called nebulae, each contai evaporating gas gl EGGs, which are t beginnings of sta dark nebulae, EG squeezed by the until they becom an EGG is suffic enough (at lea nuclear reactic becomes a sta stars such as a about 10 bill

◀ Stars are b of dust and g

Read further ▸ nebulae
▸▸ pg11 (b32); pg26 (p12)

A new star is born in our galaxy about every two weeks

1 2 3 4 5 6 7 8 9 10 11 12 13 14 15 t

It's a fact
Key statistics and extra facts on each subject provide additional information.

Photos and artworks
Illustrations and photographs accompany each caption. Diagrams are labeled to give more detailed scientific facts and information.

Binary system with much larger star

a
b
c
d
e
f
g
h
i
j
k
l
m
n
o
p
q
r
s
t
u
v
w

IT'S A FACT
...est star in the night sky is
...a relative magnitude of −1.5.

...test object in the night sky is
...but the Moon, with a relative
...e of −12.7.

...e brightest stars
...e colour of a star's light depends on
...mperature: blue stars are the
...st, red stars are the coolest.
...onomers rank the brightness of each
... with a number or 'magnitude'. The
...ightest stars have the lowest
...gnitude, which may be minus
...mbers. Some stars look brighter than
...hers only because they are closer to
...arth, so astronomers talk of 'relative'
...magnitude – the brightness of a star
...compared to others and 'absolute'
...magnitudes – how bright a star really is.

In a true binary pair, the stars turn together around their common centre of gravity

▲ Binary system with similar sized stars. The stars may be close together or millions of kilometres apart.

Twin stars
Many stars are in companion pairs called binaries. True binaries are pairs of stars that whirl round together like a pair of dancers, held together by their mutual gravity. Sometimes, one star passes in front of the other, which then seems to grow dim. Some stars look like binaries even though they are nowhere near each other, because they are in the same line of sight from the Earth. These are called optical binaries.

Record facts
Discover the superlatives within this box.

Red giants

Main sequence stars

Blue stars

The Sun

Yellow stars

White dwarfs

Red stars

Increasing brightness

Increasing temperature

▲ Graph showing how a star's brightness varies with its temperature. Medium-sized stars fall in a straight line – the main sequence – showing a simple relationship.

HOTTEST STARS

Star	Temperature
• Blue	up to 40,000°C
• Blue-white	11,000°C
• White	7500°C
• Yellow	6000°C
• Orange	5000°C

Amazing facts
Look out for facts that run along the bottom of each page.

Read further › light-years / neutron stars
▶▶ p11 (d22); pg26 (d2; k2); pg27 (b29)

Check it out!
• http://www.howstuffworks.com/star.htm

Red stars have a temperature of 2800°C; blue stars have a temperature of up to 40,000°C

20 21 22 23 24 25 26 27 28 29 30 31 32 33 34 35 36 37 38 39

Cross-references
Attached to captions and pictures are cross-references that use the unique co-ordinates grid system. These lead you to related subjects within the book.

Check it out!
Find out more by surfing the Internet.

Night sky

THE NIGHT sky is filled with thousands of points of light twinkling in the darkness. Most are stars – gigantic suns so distant they look tiny. About 2000 stars are visible to the naked eye, but there are trillions more out in space. Slightly brighter than stars are some of the planets that circle the Sun, like our Earth *(see pg18 [m15])*. Five can be seen by the naked eye: they have no light of their own, but are so near to Earth they reflect sunlight more brightly than any star shines. The brightest object of all in the night sky is also the nearest to Earth – the Moon.

IT'S A FACT

• With binoculars, you can see up to 5000 stars.

• Stars twinkle because we see them through the shimmering layers of Earth's atmosphere.

Ursa Major (the Great Bear)

Pegasus (the Winged Horse)

Patterns of stars

To find their way around the night sky, astronomers divide the sky into 88 patterns of stars, or constellations. Many constellations still have the names of the ancient Greek mythical heroes and creatures they were given long ago, such as Orion the hunter. There is no particular connection between the stars in a constellation; they just appear close together in the sky.

▶ *Four constellations visible in North America and Europe.*

The sky at night

Stars appear in the same pattern they have done for many thousands of years – although a few stars, such as Polaris, have shifted slightly since the time of the first ancient Babylonian astronomers. By studying the sky on a nightly basis, you can learn to identify bright stars, such as Sirius, and even some of the planets in our Solar System, such as Venus and Jupiter, and recognize some of the better known constellations.

▶▶ Read further > stars
pg24 (d2); pg34 (k2)

Orion (the Hunter)

Hercules

Many stars have names given to them by Arabic astronomers long ago, such as Aldebaran

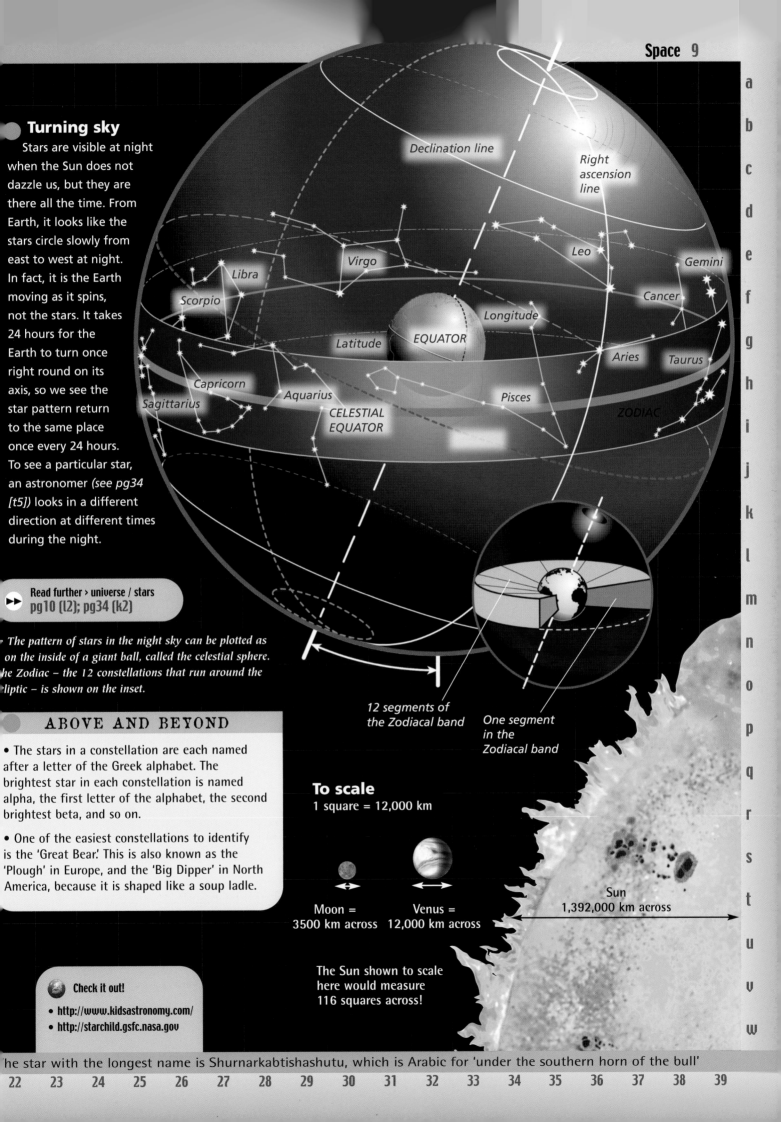

Turning sky

Stars are visible at night when the Sun does not dazzle us, but they are there all the time. From Earth, it looks like the stars circle slowly from east to west at night. In fact, it is the Earth moving as it spins, not the stars. It takes 24 hours for the Earth to turn once right round on its axis, so we see the star pattern return to the same place once every 24 hours. To see a particular star, an astronomer *(see pg34 [t5])* looks in a different direction at different times during the night.

Read further > universe / stars
pg10 [l2]; pg34 [k2]

The pattern of stars in the night sky can be plotted as on the inside of a giant ball, called the celestial sphere. The Zodiac – the 12 constellations that run around the ecliptic – is shown on the inset.

Declination line

Right ascension line

Leo

Gemini

Virgo

Cancer

Libra

Scorpio

Longitude

EQUATOR

Aries

Taurus

Latitude

Capricorn

Pisces

Aquarius

ZODIAC

Sagittarius

CELESTIAL EQUATOR

12 segments of the Zodiacal band

One segment in the Zodiacal band

ABOVE AND BEYOND

• The stars in a constellation are each named after a letter of the Greek alphabet. The brightest star in each constellation is named alpha, the first letter of the alphabet, the second brightest beta, and so on.

• One of the easiest constellations to identify is the 'Great Bear'. This is also known as the 'Plough' in Europe, and the 'Big Dipper' in North America, because it is shaped like a soup ladle.

To scale
1 square = 12,000 km

Moon = 3500 km across

Venus = 12,000 km across

Sun 1,392,000 km across

The Sun shown to scale here would measure 116 squares across!

Check it out!
• http://www.kidsastronomy.com/
• http://starchild.gsfc.nasa.gov

The star with the longest name is Shurnarkabtishashutu, which is Arabic for 'under the southern horn of the bull'

What is space?

SPACE IS everything in the universe that lies outside the Earth's atmosphere. Looking at the night sky, space seems filled with stars. Yet the distances between the stars are unimaginably vast, and there is almost nothing between them but clouds of stardust. Much of space is a vast, empty void, which is how it gets its name. No one knows how large space is, and much of it is too far away to see, but using modern technology astronomers are discovering more and more.

ABOVE AND BEYOND

• Since light takes a long time to reach Earth from distant objects in space, we see them not as they are now but as they were when the light left them. We see the bright star Deneb as it was 1800 years ago, at the time of ancient Rome.

• When we look at the Andromeda Galaxy, we see it as scientists believe it was over 2 million years ago, when the first human-type creatures evolved in Africa.

The scale of the universe

What we can see of space is only a tiny fraction of what is there. With powerful telescopes, intensely bright clusters of stars or galaxies (see pg28 [j17]) called quasars can be seen 13 billion light-years away. So if there are quasars equally far away in all directions, the universe must be at least 26 billion light-years across. The light of some stars, when seen through a telescope, may be thousands or even millions of light-years away.

►► Read further > light pg11 (d22); pg30 (j6)

IT'S A FACT

• It takes light about eight minutes to reach us from the Sun.

• It takes light four years to reach us from Proxima Centauri, the nearest star to the Sun.

▲ Even nearby stars are over 40 trillion km away; many stars are billions of times further.

Check it out!
• http://www.kidsastronomy.com/academy

If our Sun was the size of a football, its nearest star, Proxima Centauri, would be as far away as Los Angeles is from Beijing

2 3 4 5 6 7 8 9 10 11 12 13 14 15 16 17 18 19

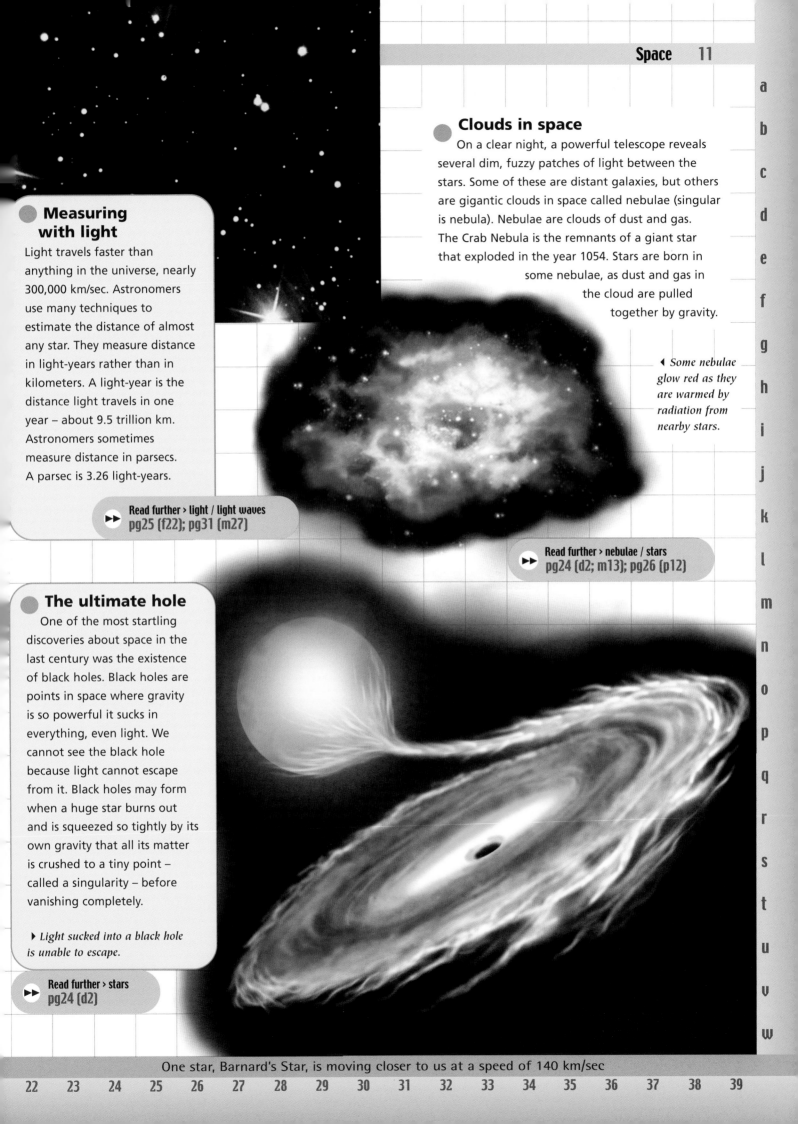

a
b
c
d
e
f
g
h
i
j
k
l
m
n
o
p
q
r
s
t
u
v
w

Clouds in space

On a clear night, a powerful telescope reveals several dim, fuzzy patches of light between the stars. Some of these are distant galaxies, but others are gigantic clouds in space called nebulae (singular is nebula). Nebulae are clouds of dust and gas. The Crab Nebula is the remnants of a giant star that exploded in the year 1054. Stars are born in some nebulae, as dust and gas in the cloud are pulled together by gravity.

◀ *Some nebulae glow red as they are warmed by radiation from nearby stars.*

Measuring with light

Light travels faster than anything in the universe, nearly 300,000 km/sec. Astronomers use many techniques to estimate the distance of almost any star. They measure distance in light-years rather than in kilometers. A light-year is the distance light travels in one year – about 9.5 trillion km. Astronomers sometimes measure distance in parsecs. A parsec is 3.26 light-years.

▶▶ **Read further › light / light waves**
pg25 (f22); pg31 (m27)

▶▶ **Read further › nebulae / stars**
pg24 (d2; m13); pg26 (p12)

The ultimate hole

One of the most startling discoveries about space in the last century was the existence of black holes. Black holes are points in space where gravity is so powerful it sucks in everything, even light. We cannot see the black hole because light cannot escape from it. Black holes may form when a huge star burns out and is squeezed so tightly by its own gravity that all its matter is crushed to a tiny point – called a singularity – before vanishing completely.

▶ *Light sucked into a black hole is unable to escape.*

▶▶ **Read further › stars**
pg24 (d2)

White ball of rock

THE MOON is the biggest, brightest object in the night sky, shining almost like a night-time sun. Yet it has no light of its own. It is just a big cold ball of rock, and it shines only because it reflects the light of the Sun. It is Earth's companion in space, about 384,000 km away, and circles slowly around it once a month. As the Moon moves around, it also rotates (turns) slowly on its axis, so that the same face always points towards us. The far side of the Moon can never be seen from the Earth's surface.

IT'S A FACT

• The Moon is barely one-quarter the size of the Earth.

• The Moon takes 27.3 days to circle the Earth, but it takes 29.53 days, or a lunar month, from one Full Moon to the next, because the Earth is also moving.

▼ *After each full Moon, the visible portion of the Moon shrinks again.*

Full Moon

Waning Moon

Half Moon

Old Moon

New Moon

Walking on the Moon

When astronauts landed on the Moon in 1969, they found a landscape of cliffs and plains, completely covered in many places by a fine white dust. This lunar dust was created long ago when the Moon's surface broke up under the impact of meteoroids. Because there is no air, wind, rain or snow on the Moon, the dust never moves – and so the footprints left behind by the astronauts will be there for millions of years.

▶ *The Earth seen from the Moon.*

Phases (changes) of the Moon

All that can be seen of the Moon from Earth is its brightly lit, sunny side *(see pg15 [c27])*. As it circles the Earth, the sunny side of the Moon is seen from different angles and so it seems to change shape. At New Moon, the Moon is positioned directly between the Earth and the Sun and all that can be seen from Earth is just a crescent-shaped glimpse of the sunny side. Over the next two weeks, more and more of the Moon is revealed until at Full Moon, when it is furthest away from the Sun, all of it becomes visible. During the next two weeks, less and less of the Moon is visible until it returns to a crescent shape, called the Old Moon.

Read further › eclipse pg15 (b22)

At night, temperatures on the Moon drop to -162°C

1   2   3   4   5   6   7   8   9   10   11   12   13   14   15   16   17   18   19

Seas and craters

All over the Moon are large, dark patches that people once thought were seas, so they are called *Mare*, from the Latin for sea. Today, scientists know they are vast, dry plains formed by ancient lava from volcanoes that erupted early in the Moon's life. Most of the craters that pit the Moon's surface also date from early in the Moon's life. They were formed by the impact of huge rocks that crashed down from space.

▶▶ Read further > Moon's surface
pg12 (o2)

▶ *The Moon's surface is pitted with ancient craters that were made by the impact of meteoroids.*

THE MOST MOONS

Planet	Moons
Uranus	21
Saturn	18
Jupiter	16
Neptune	8
Mars	2

Earth and Pluto each have 1 moon

o scale
square = 3000 km

e Earth
nd the
oon shown
scale

Moon =
3500 km
across

Earth = 12,756 km across

Moon landing

The Moon is the only other world humans have ever visited. The first men to walk on the Moon were the Americans, Neil Armstrong and Buzz Aldrin, on the *Apollo 11* space mission. They touched down on its surface on 20 July 1969. The first woman in space was the Russian, Valentina Tereshkova.

▼ *Alan Bean from Apollo 12.*

▶▶ Read further > space travel
pg12 (o2); pg32 (h32)

Check it out!

• http://kids.msfc.nasa.gov/Earth/Moon/
• http://www.dustbunny.com/afk/howdo/howdo.htm

j
k
l
m
n
o
p
q
r
s
t
u
v
w

The Moon shines partly because Moon dust contains flecks of glass that reflect the Sun

Great ball of fire

T HE SUN is a star, just like all the stars in the night sky. In fact, it is a medium-sized star in the middle of its 10 billion-year life. Yet it is much nearer than any other star – just 150 million km away. Like all stars, it is incredibly hot (*see pg25 [q23]*). Huge pressure inside the Sun creates temperatures of over 15 million°C. This tremendous heat turns the Sun's surface into a raging inferno that burns so brightly that it floodlights the Earth, giving us daylight.

(*see pg25 [q23]*)

● IT'S A FACT

• Each 6 square cm of the Sun's surface burns with the brightness of 1.5 million candles!

• The Sun is 100 times wider than the Earth.

● Inside the Sun

The Sun is made mostly of two gases: about three-quarters hydrogen and one-quarter helium. The energy that is generated inside its core takes 10 million years to rise to the surface, passing through several layers, including the glowing surface or photosphere, a forest of flames called the chromosphere and a crowning halo of fire called the corona.

▼ *The Sun's corona flares out behind the Moon during a solar eclipse.*

►► **Read further > Sun's heat** pg15 (f34)

Photosphere 6000°C

Chromosphere 10,000°C

Solar flare (10 million°C)

Convective zone

Radiating zone

Core (15 million°C)

▲ *Cutaway of the Sun showing its layers.*

Every second, the Sun releases the same energy as 100 billion hydrogen bombs

1   2   3   4   5   6   7   8   9   10   11   12   13   14   15   16   17   18   19

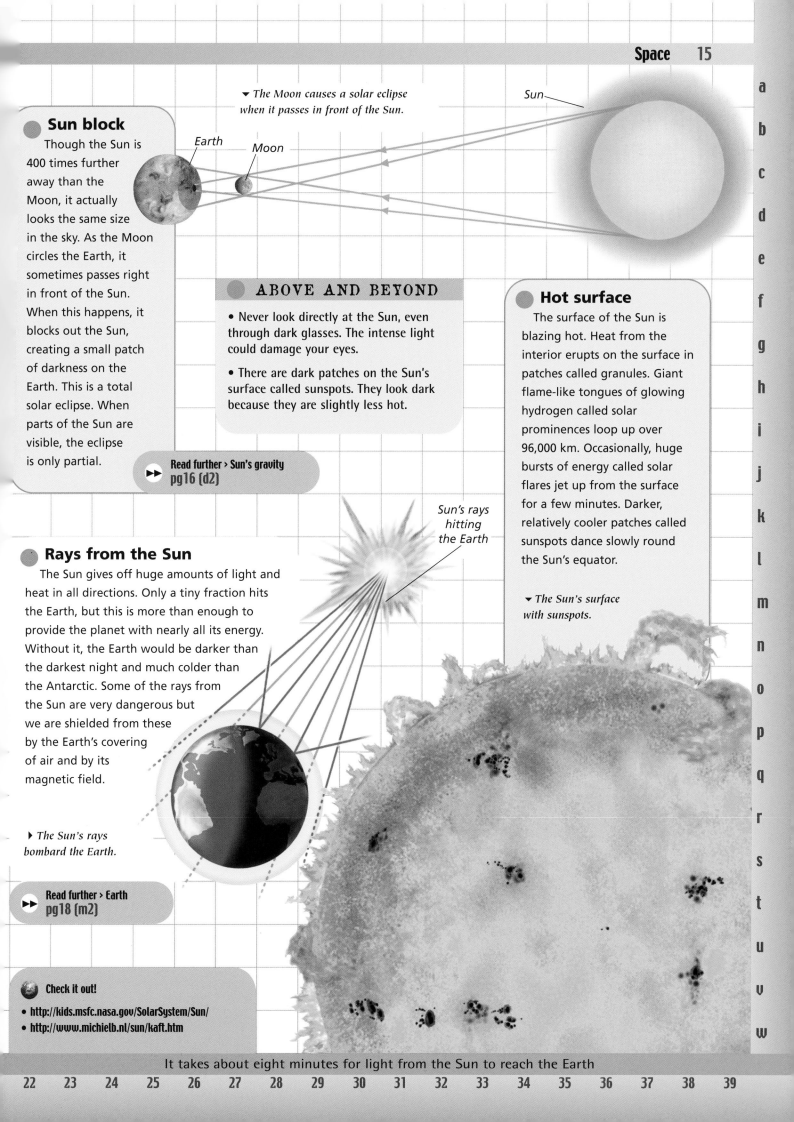

▼ The Moon causes a solar eclipse when it passes in front of the Sun.

Sun

Earth

Moon

Sun block

Though the Sun is 400 times further away than the Moon, it actually looks the same size in the sky. As the Moon circles the Earth, it sometimes passes right in front of the Sun. When this happens, it blocks out the Sun, creating a small patch of darkness on the Earth. This is a total solar eclipse. When parts of the Sun are visible, the eclipse is only partial.

▶▶ **Read further › Sun's gravity pg16 (d2)**

ABOVE AND BEYOND

• Never look directly at the Sun, even through dark glasses. The intense light could damage your eyes.

• There are dark patches on the Sun's surface called sunspots. They look dark because they are slightly less hot.

Hot surface

The surface of the Sun is blazing hot. Heat from the interior erupts on the surface in patches called granules. Giant flame-like tongues of glowing hydrogen called solar prominences loop up over 96,000 km. Occasionally, huge bursts of energy called solar flares jet up from the surface for a few minutes. Darker, relatively cooler patches called sunspots dance slowly round the Sun's equator.

▼ The Sun's surface with sunspots.

Sun's rays hitting the Earth

Rays from the Sun

The Sun gives off huge amounts of light and heat in all directions. Only a tiny fraction hits the Earth, but this is more than enough to provide the planet with nearly all its energy. Without it, the Earth would be darker than the darkest night and much colder than the Antarctic. Some of the rays from the Sun are very dangerous but we are shielded from these by the Earth's covering of air and by its magnetic field.

▶ The Sun's rays bombard the Earth.

▶▶ **Read further › Earth pg18 (m2)**

Check it out!

• http://kids.msfc.nasa.gov/SolarSystem/Sun/
• http://www.michielb.nl/sun/kaft.htm

It takes about eight minutes for light from the Sun to reach the Earth

a b c d e f g h i j k l m n o p q r s t u v w

Circling planets

EARTH IS not alone in space. Including Earth, eight planets circle, or orbit, the Sun. They move around the Sun in the same direction, in elliptical (oval) orbits, and are held in place by the pull of the Sun's gravity. Many of the other planets have their own moons and in between the planets are tiny chunks of rock called asteroids. Together, the Sun, Earth, the other planets, their moons and the asteroids are called the Solar System.

Solar System

All the planets of the Solar System orbit the Sun in the same plane. The further a planet is from the Sun, the longer it takes to complete its orbit. Mercury, the nearest planet to the Sun, takes just 88 days, Venus 225 days and Earth 365 days, but distant Neptune takes 165 years.

▶ *How the planets circle the Sun.*

Read further > solar planets
pg19 (b22); pg20 (b22)

▼ *Galaxies such as this form from swirling gas and dust.*

How it began

By measuring the age of meteorites (rocks that fall to Earth from space), scientists have worked out that the Solar System is about 4.6 billion years old. When it began to form, it was a whirling mass of stardust and gases, but as it spun around quickly, gravity began to pull it tighter together. Eventually, the dense center formed the Sun and dust further out gathered into lumps, which became the planets.

Read further > meteoroids
pg22 (i15)

A newly discovered planet orbits the star Tau Bootis in just over three days!

◀ After visiting Jupiter, Saturn, Uranus and Neptune, Voyager 2 is now leaving the Solar System.

Exploring the planets

Until barely 200 years ago, people used to think there were only five other planets in the Solar System – Mercury, Mars, Venus, Jupiter and Saturn – because only these could be seen with the naked eye. Powerful telescopes have revealed three more – first Uranus (1781), then Neptune (1846) and Pluto (1930). Now, unmanned space probes have visited all the planets except Pluto, and landed on Mars and Venus.

Read further > exploring the Solar System
pg33 (b22); pg34 (r8)

ABOVE AND BEYOND

• The size of the Solar System measures at least 20 billion km across. If Earth were the size of a grain of salt, the Solar System would be the size of a sports stadium.

• Over 70 of the known planets outside our Solar System are circling stars visible to the naked eye.

Faraway planets

Scientists guess that about 30 billion other stars in our galaxy have planets circling them, just like the Sun. Astronomers are now hunting for some of these 'extra-solar' planets. The planets are much too far away to see but they can be detected because their gravity makes their star wobble slightly. Astronomers have spotted many 'extra solar' planets – most of them as big as Jupiter. One day, they hope to spot planets as small as Earth, too. The dense atmosphere on some of the planets may give them a bright color, such as the purple shown in this artwork.

Check it out!

• http://www.stardate.org/
resources/ssguide/

a b c d e f g h k l m n o p q r s t u v w

Rocky planets

THE FOUR planets nearest to the Sun are, in order, Mercury, Venus, Earth and Mars. All four are small compared to most of the planets further out, such as Jupiter. These four planets are sometimes called the 'terrestrial' or Earth-like planets. Unlike the giant outer planets, they are made mostly of rock, and have hard surfaces on which a spacecraft could land. In fact, space probes have landed on both Venus and Mars, the nearest planets to Earth. All the rocky planets have an atmosphere (a layer) of gas – although Mercury's is barely existent – but otherwise they are very different. Earth, above all, has abundant water and life, but each planet has its own unique qualities.

IT'S A FACT
- Mercury is the second smallest planet after dwarf planet Pluto.
- Mars is about one-tenth of the weight of the Earth.

● Earth

Earth is the third planet out from the Sun, about 150 million km away from it. Earth is sometimes called the 'Goldilocks' planet after the fairy story in which the little girl chooses porridge that is neither too hot nor too cold. Earth is not so close to the Sun that it is scorching hot, nor so far away that it is icy cold. It is also the only planet that has huge amounts of liquid water on its surface. This combination makes it uniquely able to support life.

Oceans and continents are clearly visible through Earth's atmosphere.

▶▶ **Read further › Earth and Sun**
pg15 (L22)

● Pluto, dwarf planet

Besides the four inner planets, there is another rocky planet, Pluto. Pluto is a dwarf planet – smaller than our Moon – and very far away, on the outer edge of the Solar System.

◀ *Like Earth, Pluto has one moon.*

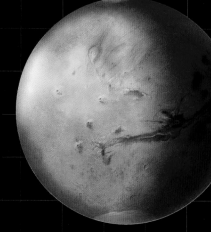

▶ *Mars' surface is cracked by a valley called the Vallis Marineris (Mariner Valley).*

Mercury

Mercury is the nearest planet of all to the Sun, often less than 58 million km away. With almost no atmosphere to protect it, temperatures on the side facing the Sun soar to 425°C while the dark side plummets to –180°C. Mercury is so close to the Sun it travels right round in just 88 days (compared to 365 days for Earth). Yet it spins very slowly, taking over 58 Earth days. So there are fewer than two days in Mercury's year.

Read further › Mercury
pg18 (d15)

The Sun scorches Mercury's surface.

▲ *Mercury has no moon.*

Venus

Venus is almost exactly the same size as the Earth. It is about 12,000 km across and weighs about one-fifth less than the Earth. Otherwise it is unlike the Earth; its atmosphere is thick with poisonous carbon dioxide and clouds of sulphuric acid. This thick atmosphere traps the Sun's heat and makes the surface a scorching desert where temperatures rise to 470°C, making it the hottest planet in the Solar System.

◀ *Venus with its thick, dense atmosphere.*

ABOVE AND BEYOND

• Like Earth, Mars has volcanoes. Olympus Mons on Mars is the biggest volcano in the Solar System, and, at 26,590 m, is three times higher than Mount Everest.

• Venus reflects sunlight so well from its atmosphere that it shines brighter in the night sky than any other star. As it appears just after sunset and just before sunrise, it is known as the evening or morning star.

Mars

Mars is the only planet with similar daytime temperatures and an atmosphere similar to Earth's, except it is mainly carbon dioxide. It is also the only other planet with water on its surface. But Mars' water is all frozen solid in ice caps, and most of the planet is a desert, with no oceans or any sign of life – just the iron-rich red rocks and dust which earn it the nickname 'red planet', as the Mars Pathfinder mission confirmed when it landed the robot exploration vehicle *Sojourner* there in 1997. Although Mars appears to be a lifeless planet, scientists hope space probes will find traces of microscopic life under the surface.

Read further › Mars
pg19 (m33); pg32 (b14)

Check it out!
• http://kids.msfc.nasa.gov/Solar System/Planets/

◀ *The Sojourner robot on Mars.*

i j k l m n o p q r s t u v w

Mars has two tiny moons called Phobos and Deimos – Phobos is 27 km across and Deimos is 15 km across

Giant balls of gas

OUT BEYOND Mars are four planets bigger than any of the others in the Solar System: Jupiter, Saturn, Uranus and Neptune. Jupiter and Saturn are especially huge. Jupiter is twice as heavy as all the other planets combined and 1300 times bigger in size than Earth! Saturn is almost as big. Despite this, all these giant planets are made mostly of gas, not rock. Only a tiny core in the center is rock. But the gas around is squeezed by the huge pressure of gravity until it is often liquid or even solid.

Jupiter

Jupiter is by far the biggest planet in the Solar System – over 140,000 km across – and it takes almost 12 years to go round the Sun. Yet despite its immense bulk, it spins round faster than any other planet. In fact, it turns right round in less than 10 hours, which means the surface is whizzing along at almost 45,000 km/h. Its surface is covered in colorful clouds of ammonia gas whipped into storm belts by violent winds, lightning flashes and thunderclaps. One storm, called the Great Red Spot, is 40,000 km across and has lasted at least 300 years. Jupiter has a faint ring system and 16 moons.

▶ Jupiter, with the Great Red Spot clearly visible, bottom left.

Check it out!

• http://www.frontiernet.net/ ~kidpower/jupiter.html
• http://www.dustbunny.com/ afk/planets/jupiter/

The surface of Neptune's moon, Triton, is the coldest place in the Solar System at -235°C

a
b
c
d
e
f
g
h
i
j
k
l
m
n
o
p
q
r
s
t
u
v
w

Saturn

Saturn is the second largest planet – a pale butterscotch-colored ball of gas over 120,000 km across. It is known as the ringed planet, because it has a spectacular halo of rings round the middle. The rings consist of countless tiny chunks of ice and rock. Although barely thicker than a house, the rings stretch out over 170,000 km into space.

▶ *Saturn is made almost entirely of gas – helium and hydrogen.*

Read further > rock core
pg20 (d15)

Neptune

Neptune is the eighth planet out from the Sun and the fourth largest planet in the Solar System. Like Uranus, Neptune is covered in deep oceans of liquid methane that fill the atmosphere with methane and give it a beautiful cobalt-blue color. It is so far from the Sun – about 4.5 billion km – that it takes 164.79 years to complete its orbit. Thus in 2011, it will complete its first orbit around the sun since being discovered in 1846.

▼ *Like Saturn and Uranus, Neptune is encircled by rings.*

Read further > wind
pg20 (d15)

Uranus

Uranus is so far from the Sun that its surface is unimaginably cold. Temperatures at the cloud tops of Uranus drop to –210°C! In this bitter cold, even the methane (natural gas) that makes up much of the planet atmosphere turns to liquid. As with Neptune, Uranus' startling blue color is caused by methane in the atmosphere.

▲ *Uranus is completely covered in deep oceans of methane.*

Read further > moons
pg23 (i33)

ABOVE AND BEYOND

• Uranus rolls around the Sun with its south pole pointing sunwards. As a result, the south pole is the hottest place on the planet. Summers here last 42 years!

• Because Jupiter is so massive, its gravity is very strong – its pressure squeezes the planet so hard that it gets hot.

Saturn is so low in density, it would float – if you could find a bucket big enough!

Space trash and hangers-on

A S WELL as nine large planets, the Solar System contains countless smaller objects *(see pg16 [k14])*. All the planets except Venus and Mercury have moons or 'satellites' circling them. Jupiter has 39. Then there are hundreds of thousands of lumps of rock, metal and ice, called asteroids. Most circle the Sun in the asteroid belt between Mars and Jupiter, and may be debris from a planet that broke up or never properly formed. Most satellites and asteroids move steadily around the Sun or their planet. But other objects, such as comets and meteoroids, hurtle round in all directions – and may even crash into planets.

Rocks from space

Most meteoroids colliding with the Earth are so small that they burn up as they enter the Earth's atmosphere. Occasionally, though, there are chunks large enough to make it all the way down to the ground. These chunks are called meteorites. Most are smaller than a fist and barely noticed. But a few are much, much bigger. These can cause havoc when they strike the ground because they are travelling at enormous speeds – not only creating giant craters on impact, but causing as much devastation as all the world's nuclear bombs if they exploded together.

◄ *Meteoroids, with a comet in the background.*

Under the ice on Jupiter's moon, Europa, there may be an ocean of water that scientists believe could support life

2   3   4   5   6   7   8   9   10   11   12   13   14   15   16   17   18   19

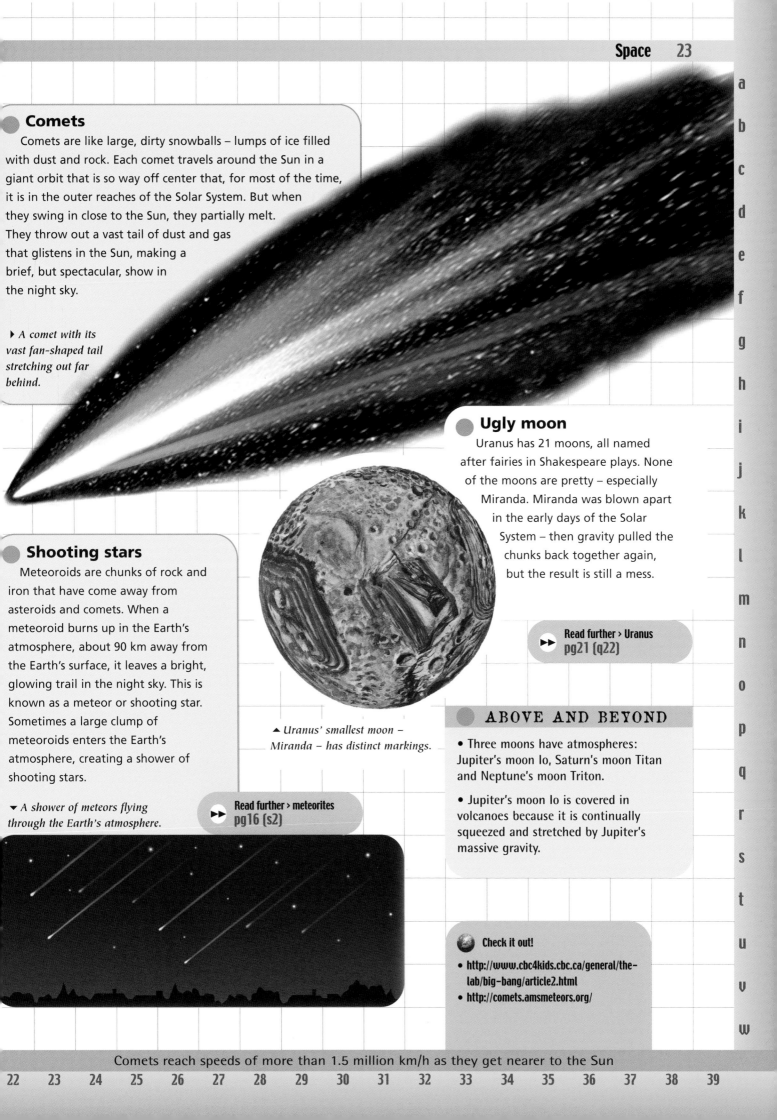

Comets

Comets are like large, dirty snowballs – lumps of ice filled with dust and rock. Each comet travels around the Sun in a giant orbit that is so way off center that, for most of the time, it is in the outer reaches of the Solar System. But when they swing in close to the Sun, they partially melt. They throw out a vast tail of dust and gas that glistens in the Sun, making a brief, but spectacular, show in the night sky.

▸ *A comet with its vast fan-shaped tail stretching out far behind.*

Ugly moon

Uranus has 21 moons, all named after fairies in Shakespeare plays. None of the moons are pretty – especially Miranda. Miranda was blown apart in the early days of the Solar System – then gravity pulled the chunks back together again, but the result is still a mess.

▸▸ **Read further › Uranus pg21 (q22)**

Shooting stars

Meteoroids are chunks of rock and iron that have come away from asteroids and comets. When a meteoroid burns up in the Earth's atmosphere, about 90 km away from the Earth's surface, it leaves a bright, glowing trail in the night sky. This is known as a meteor or shooting star. Sometimes a large clump of meteoroids enters the Earth's atmosphere, creating a shower of shooting stars.

▾ *A shower of meteors flying through the Earth's atmosphere.*

▲ *Uranus' smallest moon – Miranda – has distinct markings.*

▸▸ **Read further › meteorites pg16 (s2)**

ABOVE AND BEYOND

• Three moons have atmospheres: Jupiter's moon Io, Saturn's moon Titan and Neptune's moon Triton.

• Jupiter's moon Io is covered in volcanoes because it is continually squeezed and stretched by Jupiter's massive gravity.

Check it out!

• http://www.cbc4kids.cbc.ca/general/the-lab/big-bang/article2.html
• http://comets.amsmeteors.org/

a b c d e f g h i j k l m n o p q r s t u v w

Glowing lights

L IKE THE Sun, stars are huge, fiery balls of incredibly hot gases. They shine because they are making energy. Deep inside every shining star, huge pressure squeezes atoms of hydrogen gas together to create nuclear reactions millions of times more powerful than a nuclear bomb. These reactions boost the star's core temperature, so much that the surface glows white-hot. A star continues to glow, sending out light, heat, radio waves and other radiation until all the hydrogen is used up.

1. *Clumps of gas in a nebula start to shrink into tight balls that will become stars.*

3. Deep in its center, the new star starts making energy.

2. The gas spirals as it is pulled inwards.

4. The dust and gas are blown away and the shining star can be seen.

▲ *The four main stages of a star's formation.*

A star's life

Stars are starting up and dying down all over the universe. They begin in giant clouds of gas and dust, where material gathers into clumps called nebulae, each containing evaporating gas globules or EGGs, which are the beginnings of stars. Inside the dark nebulae, EGGs are squeezed by their own gravity until they become hot. When an EGG is sufficiently hot enough (at least 10 million°C), nuclear reactions begin and it becomes a star. Medium-sized stars such as our Sun burn for about 10 billion years.

◀ *Stars are born in clouds of dust and gas.*

Star quality

Stars create their energy in the same way as nuclear bombs, but they rarely explode. Medium-sized stars burn steadily for billions of years because there is a balance – between the heat energy, which pushes gases outward as they expand, and gravity, which pulls them back in. Only when the star's nuclear fuel burns out, is this balance broken, and the star shrinks or, on some occasions, explodes.

◀ *A medium-sized star.*

Read further › nebulae
▶▶ pg11 (b32); pg26 (p12)

A new star is born in our galaxy about every two weeks

1   2   3   4   5   6   7   8   9   10   11   12   13   14   15   16   17   18   19

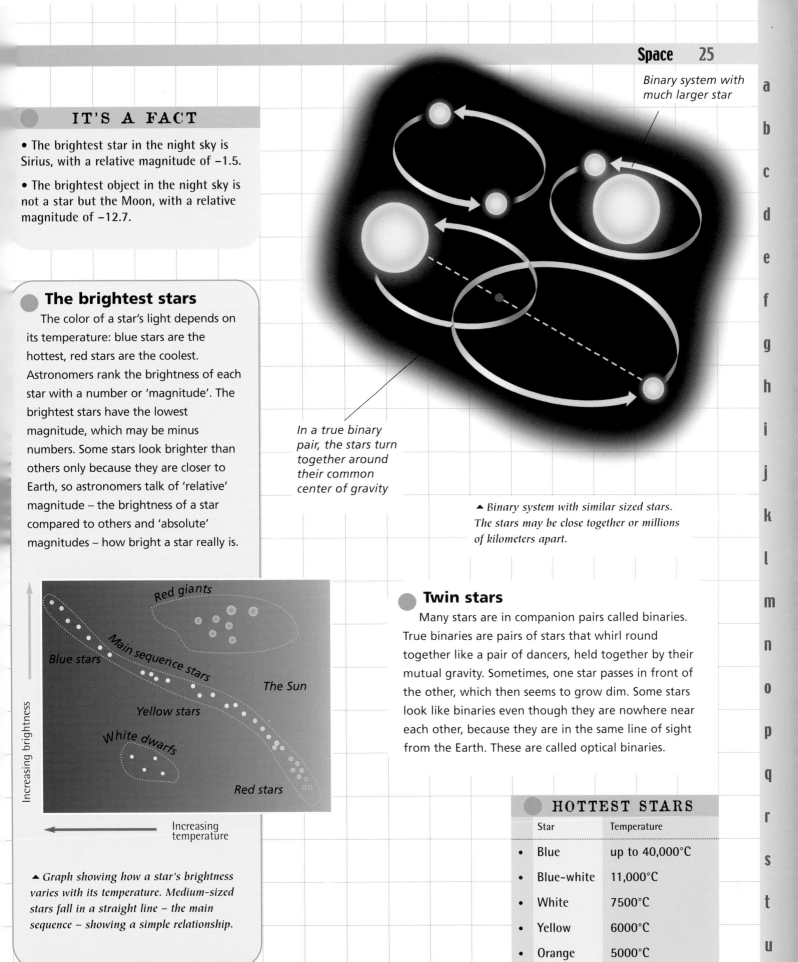

Binary system with much larger star

In a true binary pair, the stars turn together around their common center of gravity

▲ Binary system with similar sized stars. The stars may be close together or millions of kilometers apart.

IT'S A FACT

• The brightest star in the night sky is Sirius, with a relative magnitude of –1.5.

• The brightest object in the night sky is not a star but the Moon, with a relative magnitude of –12.7.

The brightest stars

The color of a star's light depends on its temperature: blue stars are the hottest, red stars are the coolest. Astronomers rank the brightness of each star with a number or 'magnitude'. The brightest stars have the lowest magnitude, which may be minus numbers. Some stars look brighter than others only because they are closer to Earth, so astronomers talk of 'relative' magnitude – the brightness of a star compared to others and 'absolute' magnitudes – how bright a star really is.

Red giants

Main sequence stars

Blue stars

Yellow stars

The Sun

White dwarfs

Red stars

Increasing brightness

Increasing temperature

▲ Graph showing how a star's brightness varies with its temperature. Medium-sized stars fall in a straight line – the main sequence – showing a simple relationship.

Twin stars

Many stars are in companion pairs called binaries. True binaries are pairs of stars that whirl round together like a pair of dancers, held together by their mutual gravity. Sometimes, one star passes in front of the other, which then seems to grow dim. Some stars look like binaries even though they are nowhere near each other, because they are in the same line of sight from the Earth. These are called optical binaries.

HOTTEST STARS

Star	Temperature
• Blue	up to 40,000°C
• Blue-white	11,000°C
• White	7500°C
• Yellow	6000°C
• Orange	5000°C

Read further › light-years / neutron stars
p11 (d22); pg26 (d2; k2); pg27 (b29)

🌐 **Check it out!**
• http://www.howstuffworks.com/star.htm

Red stars have a temperature of 2800°C; blue stars have a temperature of up to 40,000°C

a b c d e f g h i j k l m n o p q r s t u v w

Giants and dwarfs

THE SUN may be huge compared to the Earth, but it is only a medium-sized star. 'Red giant' stars are 20 to 100 times as big and 'supergiants,' such as Betelgeuse, are 500 times as big. The biggest known star is the 'hypergiant' Cygnus OB2 No.12, which is 810,000 times as bright as the Sun. Such stars burn fiercely but briefly, often lasting less than 10 million years. There are also smaller stars, such as white dwarfs, no bigger than the Earth, and neutron stars, about 20 km across. Neutron stars are the remnants of old stars that have collapsed under the force of their own gravity.

IT'S A FACT
• The night's brightest star, Sirius, also known as the dog star, has a white dwarf companion called the pup star.
• Black dwarf stars are very small, cold, dead stars. They do not give out any light.

Red giants
Red giants, such as Mira, are old stars that cool to red heat as their nuclear fuel runs out (see pg25 [q28]). At the same time they swell up to 100 times their original size. The biggest stars swell even more, until they become supergiants. Pressure in the heart of a supergiant is so huge that it may be enough to squeeze carbon atoms together to make iron – and this is probably where all the iron that exists in the universe was made. Yet despite their huge size, they have no more matter than they ever did. So even a red giant, on average, is no denser than water on Earth.

Read further > star birth
pg24 (m13)

▼ Red giant stars are 0 to 100 times as bright as the Sun.

Ring of cloud
When a giant star burns out all its fuel, it begins to collapse under the power of its own gravity. As it does, it emits huge clouds of gas at enormous speeds. These form a gigantic ring around the star that lasts a few thousand years, and eventually begins to glow as it is heated by the dying embers. This cloud ring is called a planetary nebula. It has nothing to do with planets, but got its name because it looked like a ring of planets. Counting planetary nebulae is a way of working out the brightness of a distant galaxy.

▲ Planetary nebula formed round a giant star.

Read further > medium-sized stars
pg24 (o2)

Super-dense stars

Neutron stars are the smallest, densest stars of all. They form when a star a little bigger than the Sun *(see pg14 [n13])* burns out and collapses under its own gravity. Most neutron stars are just 20 km across yet weigh as much as the Sun! A tablespoon of neutron star would weigh about 10 billion tons!

▲ *Neutron stars have a solid crust made from iron and other elements.*

Read further › black holes
pg11 (m22)

ABOVE AND BEYOND

• Neutron stars are so-called because the atoms they are made of have been so thoroughly broken down that only neutrons, tiny particles from the very center of an atom, are left.

• The star Omicron-2 (also known as 40 Eridani) is one of only a few dwarf stars that can be seen with the naked eye.

Check it out!

• http://imagine.gsfc.nasa.gov/ docs/science

White dwarf

When medium-sized stars begin to burn out, they collapse to form a white dwarf star. A white dwarf is a hot ember, little bigger than the Earth, but it is still bright. As it contains most of the matter of the original star, it is also immensely dense.

Read further › star birth
p24 (m13)

▲ *White dwarf stars are the final stage in the life of a medium-sized star.*

Supernova

Once a supergiant's core turns to dense iron at the end of its life, gravity squeezes it so hard that it collapses in just a few seconds, then blows itself to bits in a gigantic explosion called a supernova (plural supernovae). Gases are thrown out thousands of kilometers in a fraction of a second and huge amounts of light, heat and X-rays radiate out. Supernovas rarely last for more than a few months, but in that brief time they can burn brighter than a billion suns.

▼ *A supernova marks the explosive end of a massive star's life.*

Giant stars are so big it takes an hour or more for even light to cross from one side to the other

Star cities

STARS ARE not scattered evenly throughout space. Instead, they cluster together in groups called galaxies, with vast stretches of completely empty space between. The three galaxies that can be seen with the naked eye look like faint blurs in the night sky, but powerful telescopes show that they contain billions of stars. Although most galaxies are too far away to be seen, astronomers estimate that there are about 100 billion in the universe. An average galaxy, such as the Milky Way, contains 100 billion stars and is about 100,000 light-years across.

IT'S A FACT
• The Milky Way Galaxy is 100,000 light-years across.

• The Sun takes 200 million years to travel once right round the middle of the Milky Way Galaxy.

▲ Our Milky Way Galaxy seen edge-on from deep space.

The Milky Way
On clear nights, far from town when there is no Moon, a faint, hazy, white band can be seen stretching right across the sky. This band is called the Milky Way. Through binoculars, it is clear that the Milky Way consists of countless stars – indeed, it is a vast cluster of over 100 billion stars. The Milky Way appears to us as a narrow band, because we are looking at it edge-on. If we could look down on it, we would see that it looks like a giant pin wheel, with a dense bulge at its center containing mostly older stars.

Star cities
The biggest galaxies are egg-shaped or elliptical and may contain as many as a trillion stars. These elliptical galaxies probably formed a very long time ago, sometime over 10 billion years, not long after the dawn of the universe (see pg16 [u16]). Elliptical galaxies are rarely alone, and tend to group together into clusters.

▲ Elliptical galaxy clusters can contain thousands of galaxies of all kinds.

Read further > egg-shaped galaxies pg29 [k27]

The Milky Way and neighboring galaxies making up the Local Group are moving through space at over 2 million km/h

1 2 3 4 5 6 7 8 9 10 11 12 13 14 15 16 17 18 19

a
b
c
d
e
f
g
h
i
j
k
l
m
n
o
p
q
r
s
t
u
v
w

Spiral galaxies

Many galaxies, such as the Milky Way, are spiral-shaped, with a dense cluster of stars at the center. They are spiral because they are spinning, and billions of stars trail from long arms as they rotate at tremendous speed. Although we feel no movement because we are 'glued' to the Earth by gravity, our Sun is being swept round by the galaxy at nearly 100 million km/h.

▶ *Spiral galaxies are said to spin like a pin wheel.*

▶▶ **Read further › Earth**
pg18 (m2)

ABOVE AND BEYOND

• Spiral galaxies may have a giant black hole at the center, which sucks in stars like water spiraling down a plug hole.

• Although the stars make spiral galaxies look like giant fried eggs, these galaxies are actually shaped more like burgers, because they are made mostly of invisible 'dark matter'. The stars are just the filling.

Shapeless galaxies

About one in ten galaxies have no obvious shape at all. Some astronomers think these irregular galaxies formed from the debris of a gigantic space collision, after two galaxies crashed into each other.

▶ *Irregular galaxies contain many young and newborn stars.*

Check it out!

• http://www.enchantedlearning.com
  /subjects/astronomy/solarsystem

▶▶ **Read further › star birth**
pg24 (m13)

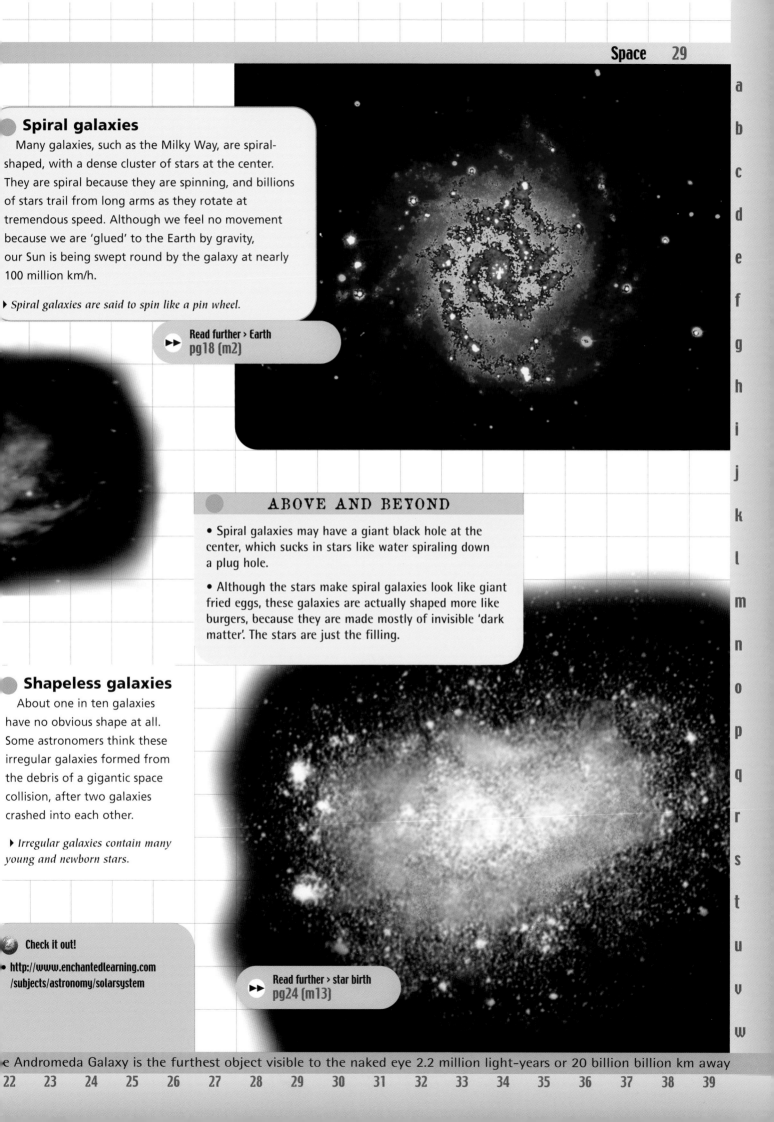

e Andromeda Galaxy is the furthest object visible to the naked eye 2.2 million light-years or 20 billion billion km away

Big Bang

THE UNIVERSE may not always have existed. Scientists think it began about 13 to 15 billion years ago with the 'Big Bang'. It is believed that one moment there was just a tiny hot ball containing everything in the universe and then a moment later, the universe burst into existence with the biggest explosion of all time, separating into basic forces such as electricity and gravity – so big that everything is still hurtling out from it – even today.

IT'S A FACT

• The furthest galaxies from Earth are hurtling away at almost the speed of light.

• The afterglow of the Big Bang can be detected in microwave radiation which emanates from all over space.

The Big Bang theory

1. At first, the entire universe was a hot ball tinier than an atom and much hotter than any star. This swelled much, much faster than the speed of light, growing to the size of a galaxy in just a tiny fraction of a second.

2. As the universe expanded, it began to cool and tiny particles of energy and matter – each of them much smaller than atoms – began to form a thick, soup-like material.

3. After about three minutes, gravity started to pull the particles together. Atoms joined togethe to make gases such as hydrogen and helium, ano the thick 'soup' began to clear and thin out. By the end of the third minute, the matter that surrounds us today had been created.

Check it out!

• http://curious.astro.cornell.edu/cosmology.php
• http://www.amnh.org/rose/hayden-bigbang.html

The universe is largely empty as nearly all the matter was obliterated early on by meeting its mirror image, anti-matt

1   2   3   4   5   6   7   8   9   10   11   12   13   14   15   16   17   18   19

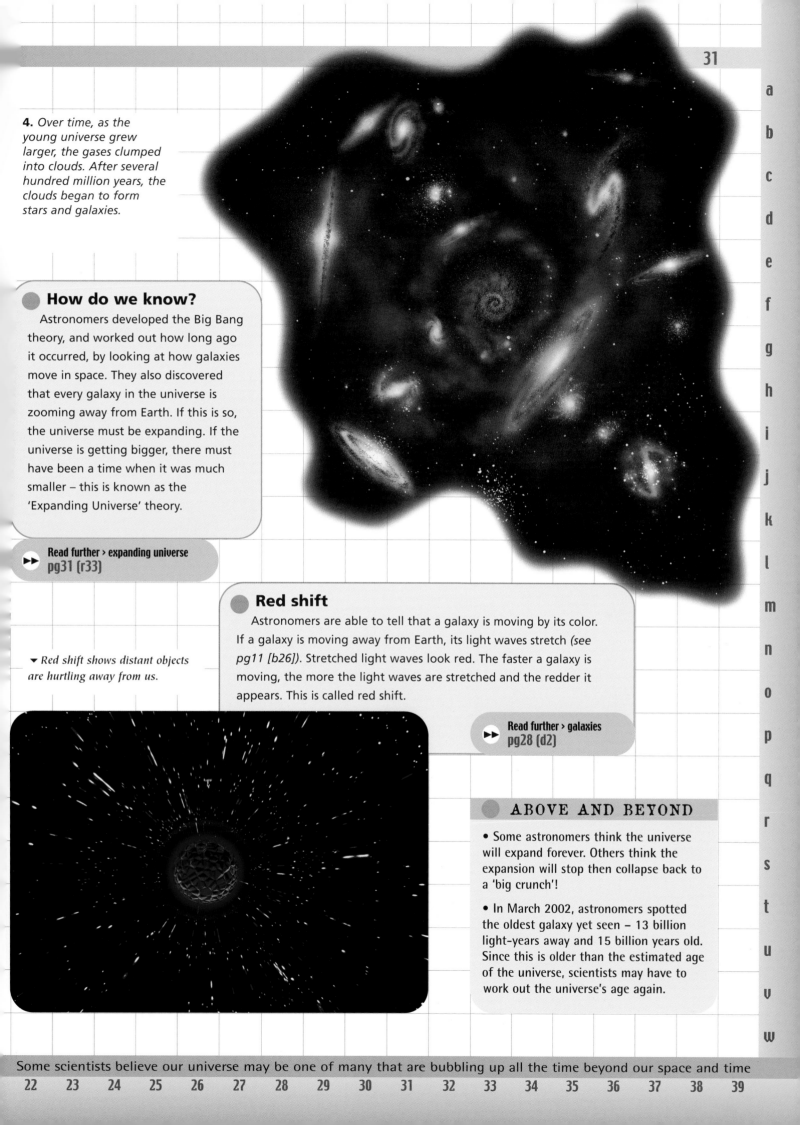

4. *Over time, as the young universe grew larger, the gases clumped into clouds. After several hundred million years, the clouds began to form stars and galaxies.*

How do we know?

Astronomers developed the Big Bang theory, and worked out how long ago it occurred, by looking at how galaxies move in space. They also discovered that every galaxy in the universe is zooming away from Earth. If this is so, the universe must be expanding. If the universe is getting bigger, there must have been a time when it was much smaller – this is known as the 'Expanding Universe' theory.

▶▶ **Read further › expanding universe**
pg31 (r33)

Red shift

Astronomers are able to tell that a galaxy is moving by its color. If a galaxy is moving away from Earth, its light waves stretch *(see pg11 [b26])*. Stretched light waves look red. The faster a galaxy is moving, the more the light waves are stretched and the redder it appears. This is called red shift.

▼ *Red shift shows distant objects are hurtling away from us.*

▶▶ **Read further › galaxies**
pg28 (d2)

ABOVE AND BEYOND

• Some astronomers think the universe will expand forever. Others think the expansion will stop then collapse back to a 'big crunch'!

• In March 2002, astronomers spotted the oldest galaxy yet seen – 13 billion light-years away and 15 billion years old. Since this is older than the estimated age of the universe, scientists may have to work out the universe's age again.

Some scientists believe our universe may be one of many that are bubbling up all the time beyond our space and time

Space travel

THE AGE of space travel dawned half a century ago, when the tiny Russian satellite *Sputnik 1* was blasted into space in 1957. Since then hundreds of spacecraft have been launched and the boundaries of space exploration are being pushed further and further, as spacecraft venture into the Solar System. In 1969, the astronauts of *Apollo 11* set foot on the Moon. In 1976, the *Viking 1* robot space probe landed on Mars. In 1973, *Pioneer 10* reached Jupiter. *Voyagers 1* and 2, launched in 1977, have flown beyond Pluto, though not yet out of the Solar System altogether.

Read further > space shuttle pg33 (h32)

ABOVE AND BEYOND

• Mars has already been visited by more space probes (see pg19 [t29]) – than any other planet, though not all were successful. It may become the first planet (beyond Earth) to be visited by humans.

• The first living creature in space was a dog called Laika, which went up in the Russian Sputnik 2 in 1957. Sadly, she could not be brought back to Earth.

Orbiter *goes into orbit around the Earth*

Orbiter *crew place satellite in space*

Space shuttle

In the early days, manned spacecraft could only be used once, with just a tiny capsule holding astronauts as they fell back to Earth. Now astronauts are carried up into orbit above the Earth by shuttle craft, which can take off and land again and again, like an airplane. The Russian version was a one-off called the Buran or 'snowstorm'; the American version is known as the space shuttle orbiter.

Orbiter *positions itself to re-enter Earth's atmosphere*

Main fuel tank falls away 130 km up

Solid fuel rocket burners fall away 45 km up

Shuttle being transported to launch pad

Orbiter *lands like a glider*

IT'S A FACT

• It will take NASA's New Horizons space probe ten years to reach Pluto – due to arrive in 2015.

• The first man in space was Russian cosmonaut, Yuri Gagarin, in April 1961.

Check it out!
• http://galileo.jpl.nasa.gov/
• http://www.nasa.gov/kids/kids_spacetravel.html

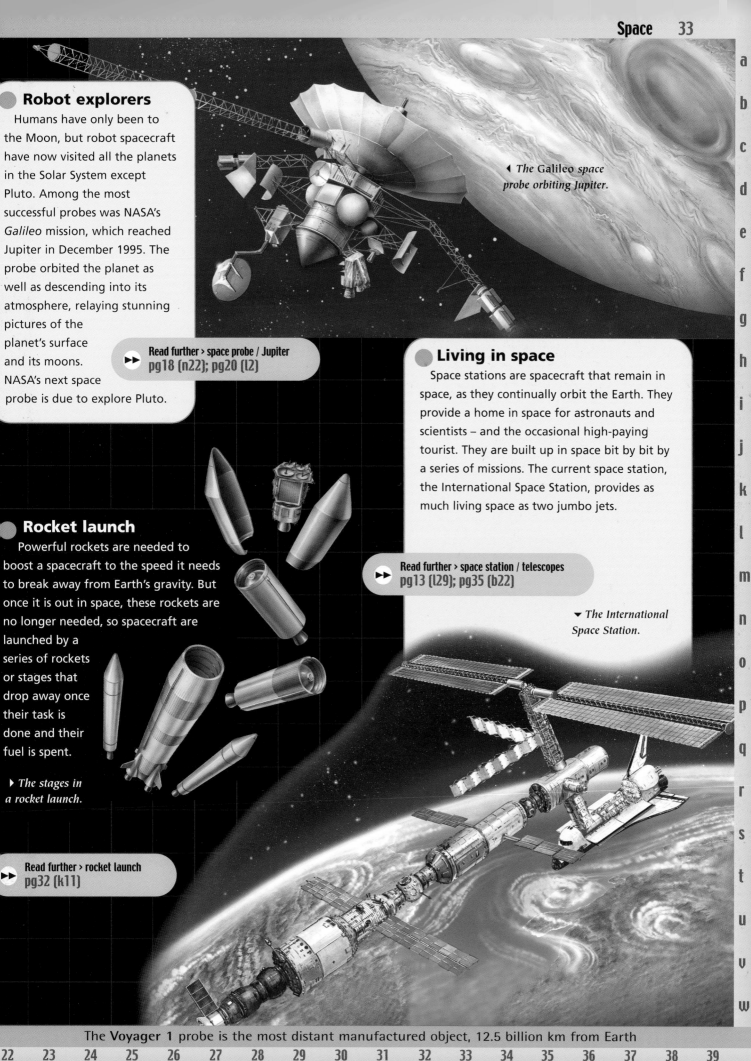

Robot explorers

Humans have only been to the Moon, but robot spacecraft have now visited all the planets in the Solar System except Pluto. Among the most successful probes was NASA's *Galileo* mission, which reached Jupiter in December 1995. The probe orbited the planet as well as descending into its atmosphere, relaying stunning pictures of the planet's surface and its moons. NASA's next space probe is due to explore Pluto.

▶▶ **Read further > space probe / Jupiter**
pg18 (n22); pg20 (l2)

◀ *The Galileo space probe orbiting Jupiter.*

Living in space

Space stations are spacecraft that remain in space, as they continually orbit the Earth. They provide a home in space for astronauts and scientists – and the occasional high-paying tourist. They are built up in space bit by bit by a series of missions. The current space station, the International Space Station, provides as much living space as two jumbo jets.

▶▶ **Read further > space station / telescopes**
pg13 (l29); pg35 (b22)

▼ *The International Space Station.*

Rocket launch

Powerful rockets are needed to boost a spacecraft to the speed it needs to break away from Earth's gravity. But once it is out in space, these rockets are no longer needed, so spacecraft are launched by a series of rockets or stages that drop away once their task is done and their fuel is spent.

▶ *The stages in a rocket launch.*

▶▶ **Read further > rocket launch**
pg32 (k11)

The **Voyager 1** probe is the most distant manufactured object, 12.5 billion km from Earth

a b c d e f g h i j k l m n o p q r s t u v w

Observing space

● **IT'S A FACT**

• The world's largest observatory complex is on top of the Mauna Kea volcano in Hawaii, at a height of 4200 m.

• The E-Merlin radio telescope array, in the UK, is so powerful that it may be able to see a bottle top from 80 km away!

UNTIL A century ago, astronomers thought all space was little more than our own Milky Way Galaxy. All they could see of the Andromeda Galaxy – the most distant object visible to the naked eye – was a fuzzy cloud. Then in the 1920s, individual stars were picked out in Andromeda for the first time, and it was clear that Andromeda was a completely separate galaxy. Astronomers began to realize that space is much, much bigger than they first thought. Now, with the aid of powerful telescopes, they can see over 50 billion other galaxies, some of which are up to 15 billion light-years away.

Observatory opening for telescope

● Star gazers

Astronomers study the sky from observatories, usually placed on mountaintops away from clouds and city lights to give a clear view of the night sky. Most observatories use either a giant receiving dish, like a large satellite TV dish, or a powerful telescope housed within a dome. As the world is constantly turning, the dish or telescope must also turn to keep track on a particular patch of sky.

▶▶ Read further › astronomers
pg8 (o2); pg34 (r8)

Dome and telescope rotate independently with Earth's rotation

Telescope

Images are displayed on computer screens

● Looking further

Telescopes are the astronomer's most valuable aid. Most telescopes work by concentrating the light from distant stars and galaxies, enabling astronomers to see objects that are far too small or dim to be seen with the eye alone. Some telescopes, called refracting telescopes, use lenses to concentrate the light. Others, called reflecting telescopes, use a curved mirror to reflect the light. Catadioptric telescopes combine the use of both lenses and mirrors.

◀ *Astronomer using a reflecting telescope.*

▶▶ Read further › astronomers / galaxies
pg11 (d22); pg28 (d2)

Telescopes in space

Looking at space through the world's atmosphere is like looking through a window of frosted glass. So astronomers put telescopes in space, on satellites orbiting the Earth, to give a clearer view. Several of these space telescopes are now working. The most famous is the Hubble Space Telescope, launched from a space shuttle in 1990. When it was launched, Hubble's main mirror was faulty but in 1993 astronauts fitted correcting mirrors.

▸ *Hubble Space Telescope.*

Read further › space stations pg33 (h32)

Seeing the invisible

Visible radiation – the light we can see – is not the only radiation beamed out by stars and galaxies *(see pg28 [i16])*. They emit invisible rays, too, such as X-rays and radio waves. These can be detected with special telescopes, revealing far more about space than would be possible with visible light alone. Radio telescopes are huge dishes that pick up natural radio signals pumped out by certain stars and galaxies. Radio astronomy allows astronomers to see right into the heart of the clouds where stars are born *(see pg24 [c18])*.

ABOVE AND BEYOND

• By linking signals from ten radio telescope dishes spread across the USA, the Very Long Baseline Array (VLBA) can pick up radio emissions from very dim and distant stars and galaxies.

• High-powered telescopes have revealed numerous tiny moons orbiting Jupiter and Saturn – some no bigger than a small town.

Read further › star birth pg24 (m13)

▲ *Radio telescopes use an array of linked dishes – the further apart they are, the clearer the images they relay.*

Check it out!
• http://skyview.gsfc.nasa.gov/
• http://www.dustbunny.com/afk/

In 2001, the Chandra X-ray telescope saw evidence of a huge black hole at the center of our Milky Way Galaxy

Glossary

Asteroid One of many thousands of large chunks of rock that orbit the Sun. The smallest are just a few hundred meters across; the biggest are more than 1000 km across. Most asteroids lie in a belt between Mars and Jupiter.

Aurora Colorful glowing lights seen in the sky in the far north and far south. They are caused by the impact of energetic particles from the Sun on the gases of Earth's atmosphere.

Big Bang The huge explosion-like expansion of the universe that may have occurred when it began some 15 billion years ago. At first, the universe was a minute, very hot ball of matter and radiation. The universe then expanded and swelled, and stars, galaxies and planets began to form.

Binary star A true binary is a pair of stars that orbit each other. An optical binary is a pair of stars that look close together in the night sky but are really far apart.

Black hole A region of space surrounding an object that is so heavy and dense that its gravity is strong enough to draw in even light. Anything falling into a black hole is crushed to oblivion. There may be black holes at the heart of spiral galaxies.

Chromosphere The lower layer of the Sun that burns at about 10,000°C, just below the photosphere. The chromosphere's color is a pale red or pink.

Comet A 'dirty snowball' of rock, dust and ice orbiting the Sun. When its orbit brings it close to the Sun, it partially melts, sending out a giant, shining tail.

Constellation A group of stars that form a pattern in the night sky. Well-known constellations include Orion and the Great Bear.

Cosmic ray Stream of radiation from the Sun.

Eclipse When one space object blocks the view of another. When the Moon blocks Earth's view of the Sun, it is a solar eclipse. When the Earth blocks the Sun's light from the Moon, it is a lunar eclipse.

Extrasolar planet A planet outside the Solar System, not circling the Sun but another star.

Galaxy A giant collection of stars in space containing millions of stars. Some are spiral shaped, some are elliptical (oval) and some are irregular in shape. The Sun is part of a local galaxy called the Milky Way Galaxy or just the Galaxy.

Gravity The force of attraction that pulls together two objects because of their mass. The heavier an object, the stronger its gravity. Every object has its own force.

Hydrogen The lightest, most common gas in the universe, and the first to form. Stars are made mostly of hydrogen and another light gas called helium.

Light-year The distance light travels in one year – about 9.5 million million km. Distances to the stars are measured in light-years.

Meteor A meteoroid that crashes into Earth's atmosphere and burns up – often seen as a glowing streak in the sky called a shooting star.

Meteorite A meteor so large that it doesn't burn up in Earth's atmosphere but instead crashes into the Earth's surface.

Meteoroid A small piece of space rock that crashes into the Earth.

Milky Way The faint band of light stretching across the night sky, made of billions of stars. This is the edge-on view of our own Galaxy, the Milky Way.

Nebula A cloud of dust and gas in space. Some glow because they contain young stars. Some glow because they reflect light from other stars.

Neutron star A very small and dense star formed when a large star explodes.

Photosphere The surface of the Sun, which is made from churning, hot gases.

Planet A large world that revolves around the Sun or another star.

Planetary nebula A gigantic ring of clouds of gas emitted at speed that form around the outside of a supernova.

Pulsar A neutron star that spins rapidly sending out regular pulses of radio waves.

Quasar A small, very distant object that looks like a star but is hundreds of times brighter than galaxies. Most quasars are billions of light-years away, the furthest things visible in the universe.

Red giant A huge red star at least ten times as big as the Sun, formed when a medium-sized star, such as the Sun, begins to burn out and swell up.

Red shift The change in color as light from galaxies turns redder because the waves are stretching away from us.

Solar System The collection of planets, moons and smaller objects that orbit (circle) the Sun.

Spiral galaxy A rotating galaxy with a spiral shape like a giant pin wheel. The Milky Way is a spiral galaxy, and the Sun is on one of its arms.

Sunspot A dark spot that develops on the surface of the Sun, which interrupts the flow of gases.

Supernova A gigantic explosion caused when a giant star runs out of energy. It throws off its outer layers and burns as bright as billions of ordinary stars.

White dwarf A small, dense star formed when a medium-sized star such as the Sun runs out of fuel.

Index

The publishers would like to thank the
following artists who have contributed to this book:
Kuo Kang Chen, Alan Hancocks, Rob Jakeway, Janos Marffy, Mike Saunders, Rudi Vizi

All other photographs are from:
Corel, DigitalSTOCK, NASA, PhotoDisc